BRITISH GEOL
Natural Environr

Geology of the country around Girvan

Explanation for 1:50 000 geological sheet 7 (Scotland)

I. B. Cameron, P. Stone and J. Smellie

First published 1986

© *Crown copyright 1986*

Bibliographical reference
Cameron, I. B., Stone, P., and Smellie, J. 1985. Geology of the country around Girvan. *Explan. 1:50 000 sheet, Br. Geol. Surv.*, Sheet 7, Scotland.

ISBN 0 11 884418 0   London   Her Majesty's Stationery Office   1986

Contents

1 Introduction 1
2 Ballantrae Complex 2
    Serpentinite and associated rocks 2
    Balcreuchan Group 4
    Intrusive igneous rocks 7
3 The Downan Point Lava Formation 9
4 The Barr, Ardmillan, Tappins and Leadhills groups 10
    Barr Group 10
    Ardmillan Group 11
    Tappins Group 13
    Leadhills Group 15
5 Silurian 16
6 Devonian 18
7 Permian 19
8 Post-Ordovician igneous intrusions 20
9 References 22
10 Glossary 25
11 BGS publications relevant to the area 28

Front cover The foreshore at Ardwell seen from the south. Rocks of the Ardwell Formation (Caradoc) are in the foreground and Girvan can be seen in the distance across the bay

Notes

National Grid references are given in the form [NX 1390 9005] or [172 198] throughout; all lie within the 100-km square NX.

Symbols in brackets after lithostratigraphic names, e.g. Balcreuchan Group (BCN), refer to symbols used on the 1:50 000 scale geological map.

Numbers at the end of figure descriptions, such as D1563, refer to photographs, both colour and black and white, in the official collection of the British Geological Survey. These photographs can be purchased as either prints or slides through the main offices of the Survey.

Authors
I. B. Cameron, BSc, P. Stone, BSc, PhD, and J. L. Smellie, BSc, PhD
British Geological Survey, Murchison House, West Mains Road, Edinburgh EH9 3LA

1 Introduction

The Girvan area, from Girvan south to Glen App, has been an area of great interest to geologists for over a hundred years. Early research in the area was carried out by Lapworth (1882) and Peach and Horne (1899), and many other workers have contributed information and opinion since then. Recently interest has been stimulated by the development of plate-tectonic theory and the appreciation that this part of Scotland occupies a unique situation in relation to major tectonic events in the Ordovician and Silurian.

The rocks are best seen on the coast between Girvan and Glen App. Exposure tends to be limited inland.

In the early Palaeozoic the area was part of the southern margin of the North American continent which lay on the north side of the Iapetus Ocean. The progressive closure of the ocean by north-westerly subduction resulted in the formation at the continental margin of an assemblage of tectonically emplaced slices of oceanic crust and sedimentary rocks. The Ordovician basic, ultrabasic and associated rocks, which make up the Ballantrae Complex, are thought to represent several fragments of oceanic crust of different origins thrust over or faulted against the continental margin. Since this central part of the Girvan area is of particular interest and complexity it is described in greater detail in the *Classical areas of British Geology* series (Stone and Smellie, in press) and it is illustrated by a separate solid geology map at a scale of 1:25 000.

South-east of the Stinchar valley the rocks are considered to be part of an accretionary complex consisting for the most part of steeply dipping fault-bounded tracts of turbidites and hemipelagic sediments, but including the lavas around Downan Point. Each tract tends to be younger than its northern neighbour. The rocks south of the fault are of Llandeilo to Caradoc age.

Part of the Ordovician and Silurian sedimentary succession unconformably overlying the Ballantrae Complex is equivalent in age to the turbidites south of the Stinchar Valley Fault. It shows evidence of a northerly or north-westerly transgression across an eroded surface of the deformed and faulted rocks of the Ballantrae Complex. There is also evidence of fault movement during deposition, with downthrow to the south-east and later overthrusting to the north-west.

The various elements making up the structure of the Lower Palaeozoic rocks are a consequence of the north-west to south-east tectonic shortening, caused by the convergence of the lithospheric plates and ultimate continental collision, during closure of the Iapetus Ocean in late Silurian or Devonian times. The strike of the rocks, which are usually steeply inclined, and most of the structures have a north-east to south-west orientation. Several lineaments which developed early in the geological history of the area persisted and were reactivated, possibly more than once. The Glen App Fault and the Stinchar Valley Fault both have net downthrows to the south-east and are the major faults in the southern part of the area. They extend north-eastwards out of the area and are part of the group of the faults, which includes the Southern Upland Fault, at the southern margin of the Midland Valley graben.

Outcrops of Devonian strata occur in the northern part of the area. The Devonian rocks rest unconformably on an irregular surface of deformed and eroded Lower Palaeozoic rocks. The original extent of Devonian deposition is unknown, but a more extensive cover to the south of the present-day outcrops may have been removed by erosion following the post-Lower Devonian faulting as a result of which the Midland Valley graben was formed. Gentle folding or tilting has affected the Devonian strata.

An outcrop of strata of Permian or possibly Lower Triassic age occurs north of Ballantrae. It is probably unconformable on the Ballantrae Complex and marks the edge of the Permo-Triassic outcrop which extends under the Firth of Clyde to Arran and Northern Ireland. Subsequent faulting and tilting gave the outcrop its north-easterly dip and truncation against the complex.

At the present time there is no quarrying or mining in the area. Formerly greywacke was quarried near Carlock [NX 095 770] for roadstone, limestone for agriculture at Aldons [NX 196 896], Craigneil [NX 147 853] and Bougang [NX 113 855], and red sandstone near Chapledonan [NS 194 005] for building stone. Best known of the rocks of the district is the Ailsa Craig microgranite, which at one time was used for the manufacture of curling stones.

There were two trials for copper south of Girvan, at Balkeachy [NX 183 933] and on the eastern flank of Byne Hill [at NX 182 945], but in neither was copper recovered in useful quantities.

2 Ballantrae Complex

The Ballantrae Complex consists of a tectonic assemblage of ultramafic and mafic igneous rocks, basaltic volcanic rocks and associated sedimentary rocks (the Balcreuchan Group), and a number of intrusions. An Arenig age for the complex is proved by graptolite faunas from mudstones associated with the basaltic volcanic rocks, and is inferred from radiometric measurements.

SERPENTINITE AND ASSOCIATED BASIC IGNEOUS AND METAMORPHIC ROCKS

Ultramafic rocks crop out in two major and a number of smaller zones within the Ballantrae Complex. The two major zones, the northern and southern serpentinite belts, form the greatest part of the total outcrop and occur as elongate south-west–north-east tracts from Lendalfoot to Byne Hill [NX 180 945] in the north and from Ballantrae to Millenderdale [NX 172 905] in the south. Several smaller outcrops are contained in the Stinchar Valley Fault zone and a further series of small outcrops forms a fault-bounded strip intervening between the northern and southern belts. Of these the most interesting, petrologically, is near Knockormal [NX 137 886].

The serpentinised ultramafic rocks, mainly harzburgite and dunite, are normally black or dark green, but locally the rock is brown and mottled dark red owing to the presence of disseminated hematite. There are also pale green serpentine veins. The serpentine in the altered ultramafic rocks is formed by the replacement or conversion of olivine and orthopyroxene, the latter recognisable as brown bastite pseudomorphs.

Although the alteration is extensive it is possible to recognise differences in the petrography of the original ultramafic parent in different parts of the outcrop. There are also indications of original magmatic cumulate and tectonic textures.

The northern serpentinite outcrop is composed mainly of serpentinised harzburgite with subordinate outcrops of pyroxenite and gabbro. The original unaltered harzburgite consisted of olivine and orthopyroxene with Cr-spinel as a common accessory mineral. The spinel is resistant to alteration and survives in the serpentinite as small crystals which in places reveal an original mineralogical layering; at one locality near Pinbain Bridge [NX 1386 9152] chromite occurs in a zone 4–5 m wide in which it forms 30 to 95% of the rock (Stone and others, in press).

Mineralogical layering is also apparent locally where there is an alternation of pyroxene-rich layers with olivine-rich layers (between 1 and 50 mm thick). Incomplete serpentinisation of the pyroxenous layers makes them more resistant to erosion so that they stand proud of the less resistant olivine-rich layers.

Bodies of pale greenish grey coarsely crystalline pyroxenite occur within the northern serpentinite and are exposed in prominent rock knolls. The pyroxenites consist mainly of varying proportions of diopside and enstatite with accessory minerals. The orthopyroxene and rare accessory olivine are invariably altered to serpentine, but the clinopyroxenes are relatively unaltered.

The pyroxenite masses can be classified on the basis of their texture into a magmatic group and a tectonised group. In most of the outcrop the rocks are coarse grained or very coarse grained with normal magmatic textures, but in a zone 0.5 km wide along the south-east margin of the outcrop the pyroxenite is generally finer grained, and there are large ragged porphyroclasts.

Within the northern ultramafic outcrop a notable pegmatitic gabbro about 5 m wide, known as Bonney's Dyke, occurs on the shore about 1 km north of Lendalfoot [at NX 1351 9108] (Balsillie, 1932). It is a pale grey, very coarse-grained rock and is not chilled at its contacts with the serpentinite and pyroxenite. It consists mainly of plagioclase and clinopyroxene largely replaced by secondary minerals and may be interpreted as tectonised gabbroic segregation of the ultramafic rocks.

At the south-east margin of the northern serpentinite a narrow zone of dynamothermally metamorphosed rocks occurs between the serpentinite and the lavas. The zone was probably originally continuous, but is now seen only on the west side of Balsalloch and Carleton Hills [NX 126 893], in the Water of Lendal around Straid Bridge [NX 1390 9005] and north-west and north of Knocklaugh Farm [NX 172 918]. The zone ranges in thickness from 20 to 200 m. It consists of epidote schists farthest from the serpentinite merging with amphibolites, garnet amphibolites and garnet metapyroxenites as the ultramafic rock is approached. The metamorphism implies a wide range of temperature and pressure, with an increase in metamorphic grade from the lavas towards the serpentinites (Jones, 1977; Spray and Williams, 1980; Treloar and others, 1980). An age of 478 ± 8 Ma (K-Ar) has been obtained from the amphibolites which dates the metamorphism (Bluck and others, 1980) believed to have resulted from the displacement of a hot ultramafic body over or against lavas of the upper oceanic crust (Church and Gayer, 1973).

A number of the fault-bounded outcrops of serpentinite in the zone between the northern and southern serpentinites are known only from geophysical anomalies or borehole core (Stone and others, 1984). However, the best-known example crops out to the north-east of Knockormal Farm, where a large loose boulder was described as an eclogite by Balsillie (1937) and further discussed by Bloxam and Allen (1960). A radiometric age of 573 ± 35 Ma (Sm-Nd) has been obtained for it (Hamilton and others, 1984). The rock type is better decribed as a garnet clinopyroxenite (Smellie and Stone, 1984) which has also been seen in situ in a nearby temporary exposure. The field relations are not clear but the occurrence has been interpreted by Smellie and Stone (1984) as a segregation within mantle harzburgite. The occurrence nearby of sheared basalts with blue amphibole fringes to green hornblendic amphibole represents rather different metamorphic conditions unrelated to the crystallisation of the 'eclogite'.

The southern serpentinite belt consists mainly of serpentinised harzburgite, but dunite and wehrlite also occur. Both intrusive and tectonic relationships between dunite and harzburgite have been recorded. In the area between Knockdaw Hill [NX 162 889], Craig Hill [NX 165 876] and Breaker Hill [NX 178 892] the rocks show mineralogical layering. In particular, layered relationships are seen between harzburgite, dunite, wehrlite and olivine gabbro. A spectacular variation within this layered sequence crops out near Poundland [NX 174 874] as dunite containing nodules up to 10 mm in diameter of chrome spinel aggregates. The rocks of this area were described as layered ultramafic cumulates by Balsillie (1932) and Jones (1977) has interpreted them as a series of cumulate cyclic units comparable to the transition zone of an ophiolite. Hence, this part of the southern serpentinite may represent a higher structural level in an ophiolite sequence than the rocks of the northern serpentinite belt. There the tectonised ultramafic rocks suggest an origin at a deeper structural level.

The southern serpentinite encloses a large number of xenoliths of basic hornfels. The rocks are dolerites and gabbros which consist principally of plagioclase, hornblende and augite, but they are subdivided on textural grounds into hornblende hornfels and beerbachites. The hornblende hornfels includes non-foliated, foliated and flasered gabbros and dolerites, most of which show a low grade of thermal metamorphism. A radiometric date from a gabbro near Millenderdale [NX 172 905] gave an age of 487 ± 8 Ma (K-Ar) (Bluck and others, 1980). The beerbachites are doleritic rocks with a recrystallised granoblastic texture indicating a high grade of thermal metamorphism. The hornblende hornfels and the beerbachites have a close chemical and spatial association despite differences in their metamorphic grade. A later phase of hydrous retrograde metamorphism has affected all the xenoliths. The chemical composition of the rocks is compatible with an origin in a mid-ocean ridge environment (Jelínek and others, 1980).

The xenoliths show a variety of intrusive relationships, a wide range of metamorphic grade and have been tectonically disrupted. Various interpretations of the xenoliths have been published (Tyrrell, 1909; Balsillie, 1937; Bailey and McCallien, 1952; Bloxam, 1955; Jelínek and others, 1980).

BALCREUCHAN GROUP

The volcanic and sedimentary rocks of the Ballantrae Complex, all assigned to the Balcreuchan Group (BCN), form three main outcrops. The northernmost is the Pinbain block in the area around Pinbain Hill [NX 149 920] north of Lendalfoot. The central outcrop lies between the two serpentinite outcrops and extends north-eastwards from Bennane Head to Loch Lochton [NX 174 925]. The southern outcrop lies on the south side of the southern serpentinite outcrop between Ballantrae and Aldons Hill [NX 192 902].

The Group consists of basaltic pillow lavas, associated breccias and various interbedded sedimentary rocks. Geochemical evidence from the basalts indicate that they have several different origins (Wilkinson and Cann, 1974; Lewis, 1975; Thirwall and Bluck, 1984) and their present juxtaposition is likely to be the result of strike-slip tectonics at a continental margin (Stone, 1984).

Graptolite faunas obtained from the interbedded sedimentary rocks give a range of Arenig ages and indicate structural imbrication in places (Stone and Rushton, 1983).

Pinbain Block The Pinbain Block includes Pinbain Hill and extends from Kennedy's Pass [NX 150 933] in the north to the Pinbain Burn [NX 137 914] in the south. Exposure is generally poor inland but very good on the coast.

The block is faulted against mafic and ultramafic rocks on the southern and eastern sides, and against upper Ordovician conglomerates to the north.

The strata consist of steeply dipping massive and pillowed basic lavas and interbedded mainly volcanogenic sedimentary rocks. The dip and younging direction is for the most part to the north-west. Both porphyritic and non-porphyritic varieties of lava are present.

Their original compositions have been altered by spilitisation and burial metamorphism in the zeolite and prehnite-pumpellyite facies (Smellie, 1984).

The interbedded sedimentary rocks include breccia, lithic sandstone, tuff, siltstone, cherty mudstone and conglomerate, both oligomict and polymict with clast size ranging from pebble to boulder. There are also thin black mélange deposits with a variety of igneous clasts, but these occur in an allochthonous, fault-bounded outcrop in the south of the Pinbain Block and may be more properly included with the Bennane Head to Loch Lochton sequence.

There are conflicting indications of the conditions of deposition. The black mudstone suggests quiet undisturbed sedimentation, but the conglomerates have rounded clasts which must have originated in fluvial or shallow marine conditions. There are also hyaloclastite deposits which have characteristics of deltaic deposition (Bluck, 1982), and there is evidence of both sub-aerial weathering of lavas and sub-aerial ash-fall (Smellie, 1984).

Graptolites found in mudstone and siltstone indicate an early Arenig age (Rushton and others, in press).

Bennane Head to Loch Lochton The outcrop lies between the northern and southern serpentinite belts and extends from the coast around Bennane Head north-eastwards in a narrowing outcrop to Loch Lochton. The strata are best exposed on the coast from Bennane Head to Games Loup [NX 103 880], on Balsalloch and Carleton Hills [NX 125 890], and on Knockormal [NX 132 881] and Knockdaw Hills [NX 163 889]. The outcrop is an extensively faulted tectonic assemblage of various volcano-sedimentary sequences and encloses several structural inliers of ultramafic rocks.

The strata consist mainly of basic lavas interbedded in varying proportions with lava breccia, black mudstone, tuff, chert, red mudstone, conglomerate and rare limestone. The lavas are mainly non-porphyritic, but porphyritic varieties are locally present. They occur as massive flows and as pillow lavas, usually vesicular, and are associated with disintegration breccias (called 'macadam' breccias by Lewis and Bloxam (1980)). The breccias contain a few fragments of chert but otherwise are composed of angular to subrounded lava fragments ranging in size from 5 to 40 mm with little matrix. Locally they enclose entire pillows.

The interbedded sedimentary rocks form a minor proportion of the sequence, but fossils have been obtained which date the sequence as early and middle Arenig and give proof of tectonic imbrication (Stone and Rushton, 1983). Many of the rocks show evidence of slumping, and locally the rock is a mélange containing exotic schist and amphibolite clasts.

The lavas from many localities are spilitised and altered by burial metamorphism in the prehnite-pumpellyite facies (Smellie, 1984). Sheared lavas containing sparse blue amphibole occur locally in the vicinity of the Knockormal [NX 136 886] ultramafic inlier (Bloxam and Allen, 1960).

The outcrop can be divided for ease of description into four.

1   Carleton and Balsalloch Hills to Loch Lochton. This area lies parallel and adjacent to the southern marginal fault of the northern serpentinite belt and includes the dynamothermal metamorphic zone or 'sole' which occurs along the fault in this area. It is limited to the south-east side by the fault which runs north-eastwards from Burnfoot [NX 108 882] past the north side of the Knockormal ultramafic outcrop.

The strata strike north-east–south-west, are steeply dipping, and young consistently to the north-west.

They are principally lavas and disintegration breccias interbedded with black siliceous mudstone, tuff, chert, red mudstone and shaley limestone. The proportion of interbedded sedimentary material increases to the north-west and there is evidence of slumping and soft-sediment disruption. The north-eastern part of this area consists of a black and grey, generally sheared, mudstone mélange containing clasts of lava, white limestone and a variety of ophiolitic rock types.

Inconclusive geochemical evidence has led

to two views of the origin of the volcanic rocks. An island-arc origin is accepted by some workers (Lewis, 1975; Lewis and Bloxam, 1977), but Jones (1977) preferred an origin as oceanic crust formed at a spreading ridge.

2 Games Loup and Troax [NX 110 875] area. The rocks of this area are best seen on the coast between Games Loup and Balcreuchan Port [NX 098 875]. At Games Loup they are faulted against the ultramafic rocks of the northern serpentinite belt and a short distance north-west of Balcreuchan Port they are separated by a major fault from the Bennane Head sequence. The succession is steeply inclined and youngs to the north-west.

The rocks consist of basaltic pillow lavas with small phenocrysts of augite. On the coast the sequence contains little breccia or interbedded sediment, but the proportion of sediment increases inland (down succession) with lava breccia reddened adjacent to the thrust contact with the Balsalloch serpentinite mass, which limits this area to the east.

The evidence for the age of the rocks in this area is an isograptid fauna of late Arenig age obtained from a borehole near North Ballaird farm [NX 122 877] (Stone and Strachan, 1981; Stone and Rushton, 1983). The bore penetrated an unexposed fault slice consisting of clastic sedimentary rocks and breccia overlying silicified serpentinite. The relationship of this sequence to the coastal exposures is unknown and the graptolite fauna is not known elsewhere in the Girvan area.

Geochemical evidence indicates that the lavas, dated by Thirwall and Bluck (1984) at 476 ± 14 Ma, have a probable origin as island-arc tholeiites (Wilkinson and Cann, 1974; Lewis, 1975; Lewis and Bloxam, 1977).

3 South Ballaird [NX 122 873] to Knockdaw Hill [NX 162 889]. This area is a north-east–south-west strip south of the Balsalloch and Knockormal ultramafic inliers, and lies east of the Bennane Head outcrop. Exposure is generally poor. For the most part the dip of the strata is steep and they young to the north-west, but there is a reversal of younging direction in the South Ballaird area which could possibly be a result of slumping.

The rocks consist mainly of non-porphyritic pillow lavas interbedded with breccia and various unfossiliferous sedimentary rocks including chert, mudstone, tuffaceous sandstone and limestone.

In the area around South Ballaird and Lochton Hill the association of tuff, black siliceous mudstone and conglomerate has been described as an olistostrome (Church and Gayer, 1973; Bluck, 1978). The conglomerates contain subrounded clasts of a variety of lithologies in a heterogeneous sandy matrix.

More than one opinion on the source of the lavas has been expressed (Lewis, 1975; Lewis and Bloxam, 1977; Jones, 1977) involving island arc, oceanic island and spreading ridge origins.

4 Bennane Head. The Bennane Head sequence is well exposed on the coast. The strike of the steeply inclined strata is north–south except south of Bennane Head where it swings round to east-west; the rocks consistently young to the west in the north, and to the south and south-west south of Bennane Head.

Lavas and breccia dominate the succession with interbedded chert and clastic sedimentary rocks which contain graptolites. In the north, chert and clastic sedimentary rocks underlie a sequence of reddened porphyritic pillow lavas with thin cherty horizons which are in turn overlain by non-porphyritic unreddened lavas with lava breccia. Conglomerate, tuffaceous sandstone and shale with graptolites overlie the lavas. Farther south the proportion of lava breccia increases and at Bennane Head the outcrop consists of lava breccia with interbeds of water-laid basaltic sandstone.

A study of the graptolites, which range in age from earliest to middle Arenig, indicates that the sequence is not a single conformable succession but has been repeated by tectonic imbrication on planes more or less parallel to bedding (Stone and Rushton, 1983).

At the southern end of the outcrop near Bennane Cave [NX 0916 8618] there are bedded cherts several hundred metres thick with rare beds of basaltic sandstone. The cherts become progressively more disturbed both tectonically and by slumping towards the top of the exposed sequence. Conglomerates are interbedded with the slumped cherts and probably represent a mass-flow deposit.

Geochemical studies favour an oceanic island origin for the Bennane Head lavas (Wilkinson and Cann, 1974; Lewis, 1975; Lewis and Bloxam, 1977; Jones, 1977).

Ballantrae to Aldons Hill The Ballantrae to Aldons outcrop lies on the north-western side of the Stinchar Valley.

In the north-east, between Craig Hill [NX 165 874] and Aldons Hill the outcrop consists mainly of crudely pillowed basic lavas, with several thin conformable chert horizons and a breccia named the Craig Hill Breccia Formation (Cil). The lavas are mostly non-porphyritic and spilitised, and have undergone burial metamorphism to the prehnite-pumpellyite facies (Smellie, 1984). The Craig Hill Breccia Formation consists mainly of basalt clasts and other basic igneous types but it also includes abundant clasts of coarse-grained acid igneous lithologies. The provenance of these is uncertain but similar lithologies do crop out at Byne Hill, a little to the north.

No fossils have been found by which these rocks can be dated.

Opinion as to the original source of the lavas is divided between an island arc (Lewis, 1975; Lewis and Bloxam, 1977) and an oceanic ridge (Jones, 1977).

Farther to the south-west around Knockdolian [NX 113 848] and Sallochan Hill [NX 128 848] there is a sequence comparable to that at Bennane Head. Porphyritic lavas are overlain by non-porphyritic lavas near Knockdolian and are succeeded by lava breccias. The breccias make up most of the outcrop in the area. They are described as 'macadam-breccias' by Lewis and Bloxam (1980) and are believed to be water-laid autoclastic breccias. Black shale and chert are interbedded with the lavas on the southern slope of Knockdolian, but no fossils have been found.

The lavas at the base of the succession have undergone burial metamorphism to the prehnite-pumpellyite facies, but the mineralogy of the overlying breccias is indicative of the lower-grade, zeolite facies of burial metamorphism.

There are two opinions on the petrogenesis of the volcanic rocks: an island-arc origin is favoured by Lewis (1975) but Jones (1977) favoured an oceanic-island origin.

The Mains Hill Agglomerate Formation (MHF) crops out between Ballantrae and Balnowlart [NX 110 837] and is markedly different from the rest of the Balcreuchan Group. The rocks consist mainly of agglomerate and tuff with isolated irregular masses of lava. The pyroclastic rocks are characterised by the presence of phenocrysts of augite in the lava clasts, and conspicuous augite crystals also occur in the matrix. Indications of bedding are scarce in the outcrop but interbedded fine-grained tuff and red chert indicate, locally at least, a north-east–south-west strike and steep dips.

At the western end of the outcrop the rocks appear to have undergone burial metamorphism to the prehnite-pumpellyite facies, but at the north-eastern end they are in the lower-grade zeolite facies. Their porphyritic character impedes attempts to determine the origin of the lavas on chemical grounds, but Thirwall and Bluck (1984) favour eruption in an island-arc environment and give an age determination of 501 ± 2 Ma.

INTRUSIVE IGNEOUS ROCKS

The Ballantrae Complex includes numerous igneous intrusions which are younger than the other elements of the complex, but are overlain unconformably by middle Ordovician sedimentary rocks. Most are dolerite and gabbro but the Grey Hill-Byne Hill trondhjemite also comes into this category.

The intrusions are most common in the northern serpentinite, where steeply dipping sheets or lenses of dolerite tend to be resistant to erosion: they form rock knolls inland and prominent upstanding masses on the foreshore between Burnfoot [NX 108 883] and Pinbain Bridge [NX 138 915]. The dolerite sheets are up to 10 m thick and have chilled margins. They have a variety of orientations but north-east and east-north-east alignments are predominant. Several dykes transect the dynamothermal metamorphic sole at the south-east margin of the northern serpentinite and intrude the Balcreuchan Group lavas to the south.

The dykes between Burnfoot and Pinbain Bridge can be subdivided into an earlier and later group. The earlier group has the geochemical characteristics of island-arc tholeiites and the later dominant group has the characteristics of intra-plate basalts (Holub and others, 1984).

Gabbros intrude the lavas and sedimentary rocks between Troax [NX 110 876] and Knockormal [NX 138 887] and the ultramafic outcrop at Balsalloch [NX 116 884]. There is evidence here of two phases of intrusion separated by an episode of faulting.

A few gabbro intrusions also occur in the southern serpentinite. They are chilled against the serpentinite and differ from the beerbachites and basic hornfels xenoliths in that they have not suffered the same degree of metamorphism. Intrusions can be seen on the north side of Knockdolian [NX 116 852], near Knockdhu Bridge [NX 132 845], at Ardstinchar Castle [NX 087 824] and at Ballantrae Harbour [NX 080 830].

The largest intrusion, forming the ridge between Grey Hill [NX 164 928] and Byne Hill [NX 181 949], varies from trondhjemite in a central elongate dome through a zone of diorite and quartz-diorite to an outer zone of gabbro and dolerite (Bloxam, 1968). The trondhjemite consists mostly of albite and quartz. The mass as a whole is intruded into, and is chilled against, serpentinite. The trondhjemite has been dated radiometrically at 483 ± 4 Ma (U-Pb) (Bluck and others, 1980).

Many of the dolerite intrusions within the ultramafic outcrop have been affected at their margins by a reaction process involving Ca-metasomatism and desilication. The process is called rodingitisation and is thought to be coeval with serpentinisation. The chilled margins of the intrusions are altered to a white or grey flinty rock, called rodingite, and are veined with white fibrous pectolite with or without prehnite (Bloxam, 1954).

Figure 1 Downan Point Lava Formation (age uncertain), on the coast south of Downan Point displays steeply inclined overturned pillow lava with the youngest rocks on the left of the picture. Concentric bands of vesicles are to be seen in the upper part of many pillows, and the bottom surfaces are moulded to the shape of the underlying pillows. (D1572)

3   The Downan Point Lava Formation

The outcrop of the Downan Point Lava Formation (DPF) lies on the south side of the Stinchar Valley Fault. It is about 1.5 km wide near Ballantrae, but it narrows between converging faults to the north-east and the outcrop pinches out near Colmonell. The main outcrop of lavas is bounded on the south-east by the Dove Cove Fault, but small lava outcrops occur to the south on the coast northwest of Currarie Farm [NX 066 785] in Currarie Glen [NX 063 783] and at Portandea [NX 046 754]. The formation is about 1 km thick but that estimate takes no account of the possible effect of strike faulting.

The rocks present are pillow lavas (Figure 1) with minor amounts of massive lava and some lenticular bands of basaltic breccia and rare thin lenticular masses of red chert and limestone. The pillow form of the lavas is well developed in many places and is particularly well seen on the coast opposite Downan Farm [NX 076 807]. The lavas are steeply dipping to the north-west or south-east and throughout the outcrop they young consistently to the north-west.

The lavas are mostly spilitised tholeiitic basalts, texturally rather variable but mainly fine grained, non-porphyritic and vesicular, and composed principally of sericitised plagioclase, clinopyroxene and variable amounts of secondary carbonates, chlorite, epidote and prehnite. Some porphyritic varieties occur with small plagioclase and rarely clinopyroxene phenocrysts. A conspicuously porphyritic lava, with feldspar phenocrysts up to 5 mm long almost wholly pseudomorphed by sericite and carbonates, occurs immediately below the Currarie Formation west of Currarie Farm [at NX 060 786] and in Shallochwreck Burn [NX 063 776].

Geochemical studies of the basalts by several authors (Wilkinson and Cann, 1974; Lewis and Bloxam, 1977) all indicate an oceanic island origin for the Downan Point lavas.

The age of the lavas is uncertain. They are separated by the Stinchar Valley Fault from the Arenig lavas and volcanogenic sedimentary rocks farther north and Walton (1961) suggested that they might be younger. The overlying rocks contain graptolites of Llandeilo–Caradoc age, which sets a minimum age for the lavas.

4 The Barr, Ardmillan, Tappins and Leadhills groups

The Llanvirn to Ashgill sedimentary rocks include the Barr and Ardmillan groups north of the Stinchar Valley Fault, the Tappins Group, between the Stinchar Valley Fault and the Glen App Fault, and the Leadhills Group, south-east of the Glen App Fault.

The strata on either side of the Stinchar Valley Fault overlap in age, but there is a marked lithological contrast between north and south. In the north the strata were deposited close to a continental margin, but south of the fault the rocks consist mainly of turbidites and hemipelagic deposits. The contrast implies that there has been either considerable tectonic shortening across the Stinchar Valley Fault or a large lateral displacement along it.

BARR GROUP

The Barr Group rests unconformably on the Ballantrae Complex and is transgressive north-westwards across it. The main outcrops consist of a fault slice in the Stinchar valley, an arcuate strip around Aldons Hills westwards to Knockbain [NX 163 900], a tract from Kennedy's Pass to Byne Hill, and an outcrop at the eastern edge of the map from Laggan Hill [NX 202 948] south to Laigh Letterpin [NX 202 927]. The rocks are interpreted as an upper slope and shelf sequence. Synsedimentary faulting with south-easterly downthrow has also exerted an influence on sediment distribution and thickness.

The stratigraphy of the Barr Group has been described by Williams (1962) and Ince (1983). The tripartite subdivision of the group adopted by Ince is used here. Williams dated the strata as Caradoc, but a subsequent revision of Ordovician biostratigraphy tends to favour a Llanvirn–Llandeilo age with the series boundary probably falling within the Stinchar Limestone Formation (Ingham, 1978). The faunas bear a closer resemblance to those of the Appalachian area than they do to those of England and Wales.

Kirkland Conglomerate Formation (KdC) The oldest formation of the Barr Group is confined to a fault slice in the Stinchar valley, but it is best developed beyond the limits of the map area to the east, where at least 250 m of clast-supported pebble conglomerate are composed mainly of basalt with less common clasts of gabbro, serpentinite and chert, and rarely of foliated granite. Sandy horizons interbedded with the conglomerate increase in number upwards and the top part of the formation, exposed in the Stinchar valley, is a fine-grained fossiliferous sandstone known as the Confinis Flags from the brachiopod *Valcourea confinis*. There is also a rich trilobite fauna (Tripp, 1962).

Stinchar Limestone Formation (SLF) The maximum development of the formation is beyond the eastern margin of the map, where it is up to 60 m thick. At Aldons [NX 197 896], where the Kirkland Formation is absent, it is only about 18 m thick and rests directly and unconformably on the Ballantrae Complex. The formation is variable in lithology, but is commonly a pale grey rubbly algal limestone, locally very fossiliferous (Williams, 1962; Tripp, 1979), with grey-green mudstone and siltstone partings. In places it includes calcareous sandstone and conglomerate.

North and west of Aldons the limestone passes laterally into a conglomerate and is overlapped in places by the Benan Conglomerate Formation. The limestone is also exposed near Colmonell in three small fault slices in the Stinchar valley.

Benan Conglomerate Formation (BnC) The main outcrops of the formation are an area between Laggan Hill and Laigh Letterpin, a tract from Byne Hill to Kennedy's Pass, an arcuate strip from Aldons westward to Knockbain, and a fault slice in the Stinchar valley. The formation is thickest at the eastern end of the Aldons-Knockbain outcrop, where it is about 640 m thick, but there is marked attenuation westwards in the same outcrop and the thickness is reduced to less than 90 m west and north-west of Aldons Hill. The formation consists of a boulder conglomerate containing matrix-supported clasts up to 1 m across. The clasts are mainly of basalt, dolerite and gabbro, but also include vein quartz, chert, greywacke and pink granite. Locally clasts of Stinchar Limestone are present. An exotic limestone clast containing fossils of Lower Canadian (Tremadoc) age has also been found in the conglomerate (Rushton and Tripp, 1979) and granite boulders have given radiometric ages of about 560 and 470 Ma (Rb Sr) (Longman and others, 1979).

The Benan Conglomerate Formation includes at its base the Superstes Mudstone Member. There are small exposures of this fossiliferous mudstone at Craigneil [NX 143 852] and at Aldons (Tripp, 1976; Tripp and others, 1981).

ARDMILLAN GROUP

The Ardmillan Group, made up of four formations ranging in age from Caradoc to Ashgill, consists mainly of turbiditic sandstone, siltstone and mudstone with conglomerate prominent in the basal part of the sequence. The lateral impersistence of the strata together with the occurrence of similar lithologies at different horizons causes difficulties in establishing the stratigraphy in the lower part of the group.

The sediments were deposited from low- and high-density turbidity currents in a neritic, pro-deltaic or submarine fan environment. Slumping has affected some beds and there is other evidence of instability.

Balclatchie Formation (Bch) The Balclatchie Formation, which is up to 300 m thick, is well exposed on the coast in the vicinity of Kennedy's Pass. It consists of blue-grey nodular mudstone, siltstone, fine sandstone and laterally impersistent conglomerate units. The thickest of these, the Kilranny Conglomerate Member (KrC) at the top of the formation, is up to 150 m thick and consists of pebbly coarse sandstone with dispersed cobbles and boulders. It becomes poorly stratified upwards.

The fauna includes brachiopods, trilobites, corals, ostracods, graptolites and conodonts. The brachiopods are diagnostic of a post-Benan age (Williams, 1962; Tripp, 1980) and the formation is dated as early Caradoc.

Ardwell Formation (Ard) The Ardwell Formation, about 750 m thick at Ardwell and up to 1400 m thick inland, consists of fine flaggy sandstone, dark greenish grey siltstone and silty mudstone with sandstone laminae, and thin beds and lenses of coarse sandstone and pebble conglomerate (Figure 2 *overleaf*). There are also thin black graptolitic shale partings.

There is much evidence of contemporaneous disturbance; in addition to convolute and contorted bedding, there are intraformational breccias, slump scars and collapse breccias.

The strata on the Ardwell shore [NX 158 940] are folded into a spectacular series of cascade folds (Figure 3 *overleaf*) (Williams, 1959; Williams and Spray, 1979). There is some doubt about the origin of these folds. They are stratigraphically confined and show features which suggest to Bluck (*in* Ingham, 1978) that they may have been formed by soft-sediment deformation rather than tectonic processes.

The part of the Craighead Inlier that lies within the district consists of strata probably equivalent in age to the Ardwell Formation.

The fauna of the Ardwell Formation includes graptolites, brachiopods, trilobites and orthocone fragments, which indicate a middle or late Caradoc age (Williams, 1962; Tripp, 1980).

Whitehouse Formation (Whe) The Whitehouse Formation, about 230 m thick, is a lithologically diverse sequence most of

Figure 2 Ardwell Formation (Caradoc), at Ardwell, south of Girvan. Fine-grained greywacke, siltstone and mudstone are interbedded. The youngest beds are on the left of the picture. (D2726)

Figure 3 Folding in the Ardwell Formation (Caradoc), Ardwell shore, 9 km south-west of Girvan. (D1563)

which is exposed on the foreshore at Whitehouse [NX 165 947]. The basal unit of the formation consists of calcareous sandstone, graded coarse to fine, pebbly at the base and interbedded with pale grey-green shaley mudstone. This is succeeded by fine-grained grey sandstone and greyish green shaly mudstone. The sandstone coarsens upwards into pebbly sandstone.

The remainder of the formation is predominantly fine grained. Striped grey mudstone and siltstone with sandstone (which are not exposed on the Whitehouse shore), are succeeded by the youngest beds in the formation, which consist mainly of green and reddish brown silty mudstone.

A fauna of trilobites that includes both blind forms (such as *Dionide, Novaspis* and agnostids) and large-eyed forms (mainly cyclopygids) occurs in the upper part of the formation; graptolites are also present. The fauna indicates a late Caradoc to early Ashgill age and, together with sedimentological evidence, suggests deposition on the unstable upper slopes of a submarine fan (Ingham, 1978).

Shalloch Formation (Slh) The Shalloch Formation is exposed on the shore from Woodland Bay [NX 170 953] south to Shalloch [NX 180 960]. It is up to 300 m thick, but much of this has been removed in places below the Silurian unconformity. It consists of a sequence of alternating fine- to medium-grained turbiditic sandstone and pale green silty mudstone. Individual sandstone beds are persistent laterally and can be identified over a wide area. Convolute bedding and sandstone dykes occur in the lower part of the formation.

Thin graded limestone beds are present at a few horizons and include an early Ashgill shelly fauna (Harper, 1984). Graptolites have also been found.

TAPPINS GROUP

The Tappins Group (TAP) outcrop lies between the Glen App Fault and the Stinchar Valley Fault (Figure 4). The thickness of the group is very difficult to estimate. Williams (1962) deduced a thickness of about 1000 m for the Dalreoch Division and it is possible that further to the south-west the sequence is up to 2000 m thick. However, it is not possible to estimate the effect of faults in poorly exposed ground and in an imperfectly known sequence.

The group consists of greywacke and shale with some conglomerate, but it includes, at the base, the Currarie Formation; a diverse association of various lithologies. It has not been possible to define other formations within the group.

Currarie Formation (CUF) The Currarie Formation occurs as numerous small fault-bounded outcrops in the area south of Downan point, between Wilson's Glen [NX 070 800] and Portandea [NX 047 754]. It is about 90 m thick north of Currarie Port but could be as much as 300 m elsewhere. The complexity of the faulting makes estimates of thickness uncertain. The formation is a heterogeneous association of red siliceous mudstone interbedded with red chert, grey green mudstone and chert, and lava breccia. The formation rests on the Downan Point Lava Formation and is overlain by the turbidites of the Tappins Group.

The red siliceous mudstone, with chert beds in places, rests on the lavas and is interbedded with and overlain by grey-green graptolitic mudstone with rare thin greywacke beds. The red mudstone contains radiolaria. Lava breccia is interbedded with the mudstone at several localities on the coast west of Currarie Farm [NX 066 785] and at

Figure 4 Tappins Group (Llandeilo-Caradoc), Finnarts Hill. Medium- and fine-grained greywacke and siltstone with plane laminations. Younger rocks are on the left of the picture. (*I. B. Cameron.*)

Portandea. It consists of fragments of amygdaloidal and vesicular lavas including whole pillows and fragments of pillows up to 1 m in diameter. Within the breccia there are irregular masses and streaks of red mudstone and grey green mudstone, and also fragments of chert and jasper. Individual beds of breccia cut across the bedding of the mudstone.

Lava fragments, including masses of pillow lava up to several tens of metres across, occur within the red siliceous mudstone on the coast north of Currarie Port [NX 055 780], at Brackness Hole [NX 051 768], and at Portandea. The occurrence of lava fragments within the mudstone ranges from a few scattered blocks in an apparently unstructured mass of mudstone to numerous blocks supported in a red mudstone matrix.

The beds of breccia are interpreted as debris flow deposits periodically and locally interrupting predominantly mudstone deposition and disrupting the sediments. The red mudstone with masses of lava has been interpreted as an olistostrome (Barrett and others 1982).

The outcrops are very much affected by faulting, and there are thin fault slices of red mudstone within the greywacke and shale sequences near Glendrissaig [NX 055 762] and Portandea. The spacing of the red mudstone outcrops indicates that the north-eastward-trending faults in this area are only 150 m to 200 m apart.

A Llandeilo age for the formation is given by graptolites found in grey-green mudstone west of Currarie Farm (Knockgowan Cliff). The fossils are tentatively referred to the lower part of the *Nemagraptus gracilis* Zone.

Tappins Group above the Currarie Formation The rest of the Tappins Group outcrop consists of greywacke, shale and conglomerate with a predominantly north-east strike. The strata are commonly highly inclined and mostly young towards the northwest.

The base of the greywacke and shale sequence rests on the Currarie Formation and the sedimentary contact can be seen on the coast [at NX 0565 7825] west of Currarie Farm.

There are variations in the lithology within the area but poor exposure prevents their being mapped out.

In that part of the outcrop north and north-east of Portandea, and south-east of the Dove Cove Fault, the rocks consists of fine- to medium-grained greywacke interbedded with shale and shaley mudstone. They are dark greenish grey to mid-grey. Flattened nodules occur in the finer beds. On the coast around Portandea and northwards to Brackness Hole, and also in the vicinity of Glenapp Castle [NX 094 807] the greywacke and shale show signs of soft-sediment disruption.

In the southern part of the outcrop, in a tract from Finnarts Point [NX 044 742] and Turf Hill [NX 052 755] north-eastwards to Glen Tig, medium- and coarse-grained greywacke is interbedded with conglomerate and coarse greywacke with matrix-supported pebbles and cobbles. Fine sandstone and siltstone beds commonly have a well developed parallel lamination and the proportion of shale in the sequence is small. The main conglomerate members are known as the Glen App Conglomerate (GAPC) south-east of Ballantrae, the Finnarts Conglomerate at Finnarts Point and the Corsewall Point Conglomerate (CPC) (Walton, 1957; Kelling, 1961).

The coast section from Finnarts Bay to Nelson's Cove has yielded graptolites indicative of the *N. gracilis* Zone and possibly the overlying *Climacograptus peltifer* Zone of early Caradoc age.

In the area east and north-east of Colmonell the Tappins Group outcrop can be subivided across faulted boundaries on the basis of lithology and colour. The larger part of the outcrop consists of dull greenish grey greywacke and shale, the former commonly containing grains of red jasper. The strata are folded into a syncline trending north-east–south-west on the south-east side of the Stinchar Valley Fault and they were called the Dalreoch Division of the Tappins Group by Williams (1962). Reddish brown and purplish grey greywacke, conglomerate and mudstone occur in faulted outcrops on the south-east side of the Stinchar valley, from Heronsford [NX 115 838] to beyond Pinwherry, and also on the north-west side of the valley [at NX 190 876] near Daljarrock.

Rocks of this lithology were assigned to the Traboyack Division by Williams (1962).

LEADHILLS GROUP

The Leadhills Group (LHG) crops out in the area to the south-east of the Glen App Fault. Much of the area is very poorly exposed with an extensive cover of mainly thin hill peat. The south-eastern part of the area has not been resurveyed and the group is not subdivided into formations.

The rocks strike north-east–south-west and except in a few localities in the south-east of the area, young to the north-west. The strata are in many places steeply inclined to the south-east. They consist of fine- to medium-grained mid-grey greywacke beds ranging in thickness from 0.2 to 1.5 m, averaging about 1 m, interbedded with silty slate and mudstone. The greywacke is flaggy bedded and slightly micaceous. The finer beds tend to be platy or laminated and are finely micaceous. A few dispersed granules and small pebbles occur in the basal part of some greywacke beds. Sole structures on the greywackes are uncommon.

Close to the Glen App Fault the rocks are crushed and sheared, and they are veined and impregnated with carbonate. Pyrite is also conspicuous in the strata close to the fault on the south-east side.

Locally there are beds of grey fissile silty shale and in Tun Glen, near Carlock House [at NX 1011 7712], there is a faulted exposure of crushed devitrified bedded cherts and mudstones.

No fossils have been obtained from the Leadhills Group within the limits of the map, but graptolite faunas have been collected from thin seams of black mudstone interbedded in the greywacke and shale sequence just south of the Glen App Fault near Finnarts Bay [NX 051 727]. The faunas indicate the presence of the *Nemagraptus gracilis* Zone of Llandeilo to early Caradoc age.

5 Silurian

Several small outcrops of Silurian strata occur in the northern part of the area. These include small coastal exposures at Woodland Bay [NX 170 954] and Craigskelly [NX 178 960] 2–3 km south of Girvan and a poorly exposed outcrop about 2 km south-east of the town. The last area is the western part of the outcrop which is well exposed in the Penwhapple Glen about 5 km east of Girvan, beyond the eastern edge of the map. The rocks have both a shelly and a graptolite fauna which indicate a Llandovery age. The strata, their ages, correlation and fossil content have been described by Cocks and Toghill (1973) and their stratigraphy is followed in this account.

The succession in the coastal exposures is divided into three formations, the Craigskelly Conglomerate (CyC), the Woodland Formation (WlF) and the Scart Grits (Sca). It rests uncomformably on Ordovician rocks of the Shalloch Formation.

The Craigskelly Conglomerate is about 40 m thick at Craigskelly, but at Woodland Point it is reduced to discontinuous pockets about 1 m thick. It consists of well rounded pebbles, cobbles and boulders, clast-supported in a grey-green coarse sand matrix. Among the clasts are green mudstone, granite, jasper, chert, felsite, dolerite, gabbro, serpentinite and quartz.

The Woodland Formation, about 21 m thick, overlies the Craigskelly Conglomerate at Craigskelly, but at Woodland Point thin lenses of the conglomerate are overlapped by the Woodland Formation which rests directly and unconformably on the Shalloch Formation. The basal part of the formation consists of fine- to medium-grained blue-grey sandy beds with thin interbeds of green siltstone and prominent calcareous lenses full of shells. These are succeeded by pale green finegrained sandstone and light to dark grey banded siltstone and shale with graptolites that indicate the upper part of the *Monograptus cyphus* Zone (Cocks and Toghill, 1973).

The Scart Grits overlie the Woodland Formation and are exposed at Woodland Point, Scart Rock [NX 171 955], and north-east of Craigskelly. At Woodland Point they are 45 m thick. The formation includes at its base the Quartz Conglomerate member, which is best developed north-east of Craigskelly where it varies in thickness from 14 to 24 m. The Quartz Conglomerate consists of conspicuous angular to subrounded clasts of white quartz and a variety of other lithologies including lavas, chert, dolerite and red jasper. The clasts are closely packed and are mainly pebbles with scattered cobbles and boulders up to 0.35 m. The base of the Quartz Conglomerate is channelled deeply into the siltstone of the Woodland Formation (Figure 5). Prominent linear grooves are present as casts on the underside of the conglomerate and the siltstone beds are thrown into tight, disharmonic folds that are clearly truncated by the conglomerate. In places, siltstone wedges project into the conglomerate, and several have been ripped free, remaining as large clasts within the overlying deposit. The evidence is interpreted as indicating the forceful and rapid deposition of the conglomerate on incompletely consolidated siltstone. The remainder of the formation consists of very coarse thick-bedded sandstone with conglomerate and sandstone.

The outcrop to the south-east of Girvan has not been remapped during the revision. It is poorly exposed and the strata are assumed to be an extension of the outcrop exposed in the Penwhapple Burn which has been described by Cocks and Toghill (1973), on whose published work the following account is based.

The sequence in the Penwhapple outcrop consists of eleven formations, which for ease of representation on the map are divided into two groups, called the Penwhapple North Group, not exposed on the sheet area, and the Penwhapple South Group (PSo). The boundary is taken at the mid-Llandovery stratigraphical break at the base of the Lower Camregan Grits.

The Penwhapple South Group includes the Tralorg Formation, the Saugh Hill Grits and the Pencleugh Shale. The Tralorg Formation lies unconformably on Ordovician rocks and consists of grey and black mudstones, shales and thin siltstones with a sparse shelly fauna at the base and graptolites representing the *M. cyphus* Zone. The formation is estimated to be about 60 m thick. The Saugh Hill Grits are about 365 m thick in this area and consist of greyish green thick-bedded turbidites with thin shales that yield graptolites of the *M. gregarius* Zone, and rare conglomeratic beds. The Pencleugh Shale is about 150 m thick and consists of grey and black graptolite shale. It is referable to the *M. convolutus* Zone.

Figure 5 Quartz Conglomerate, a member of the Scart Grits Formation (Llandovery), channelled into and disrupting the siltstones and shales of the Woodland Formation. (D3528)

6 Devonian

Outcrops of rocks assumed to be of Devonian age occur in the northern part of the district around Girvan and in a small outlier in the area of the Bynehill Burn [NX 185 945] south of Girvan. Exposures in the southern outcrop occur mainly in the Bynehill Burn and its tributaries, and in the northern outcrop exposures are limited to a small area on the coast near Chapeldonan [NS 189 003] and in the hillside east of Whitehill [NX 202 984]. A small part of the outcrop near Chapeldonan is believed on lithological evidence to be Upper Old Red Sandstone (UORS), but the remainder is assigned to the Lower Old Red Sandstone (LORS).

In the Bynehill Burn outcrop the dips are predominantly to the north-west but the strata occupy a shallow syncline which opens out to the north and is faulted along its western and southern margins, where the structure is repeated. The maximum residual thickness of strata in the outcrop is at least 150 m. The few indications of dip in the Girvan outcrop are also to the north-west, but no estimate of thickness is possible. In the Bynehill Burn outcrop the strata rest unconformably on rocks of Ordovician age and in the Girvan outcrop the Old Red Sandstone rocks are faulted against rocks of Ordovician and Silurian age.

The Lower Old Red Sandstone sequence consists of sandstone, pebbly sandstone, conglomerate and rarely silty sandy mudstone, reddish to purplish brown and locally speckled, banded or mottled pale green or buff. The sandstone is variously massive, cross-bedded or plane-bedded, mainly fine to medium grained with discontinuous laminae of coarse sand. Dark purple or green silty mudstone occurs as thin beds and also as flakes in the sandstone and pebbly sandstone. The mudstone and finer grades of sandstone are micaceous.

The conglomerate and pebbly sandstone contain rounded to subangular pebbles, cobbles and rarely boulders, mainly matrix-supported in medium-grained red sandstone. The clasts consist of sandstone, mudstone, granite, quartzite, black and green chert, jasper, lava, dolerite and greywacke.

A small part of the outcrop near Chapeldonan may be Upper Old Red Sandstone facies. The strata include a sandy cornstone and nodular reddish calcareous sandstones. The strata have a low dip to the south-west in contrast to the north-westerly dip of the Lower Old Red Sandstone.

No fossil material has been found in the Old Red Sandstone in the Girvan area.

Three flooded quarries near Chapeldonan were formerly worked for building stone.

7 Permian

The outcrop of Permian rocks occupies a narrow coastal strip 3 km long, 0.5 km wide, from Ballantrae north to Bennane Lea [NX 091 860]. The Permian strata are in faulted contact with the rocks of the Ballantrae Complex except at their northern margin, where the contact is probably an unconformity. The outcrop is continuous with the large outcrop of Permian and Triassic rocks which underlies the Firth of Clyde and the North Channel. Exposure is poor and largely restricted to the intertidal zone although there are a few exposures in the burn next to Corseclays Farm [NX 096 847].

The strata dip consistently at about 15° to the north-east so that the oldest part of the succession crops out around Ballantrae and the strata become younger northwards. The sequence is divided into the Park End Breccia Formation (Pke) at the base, succeeded by the Corseclays Sandstone Formation (Cly), which includes the thin Bennane Lea Breccia member (Bnl). The whole succession is estimated to be about 1000 m thick.

The Park End Breccia Formation, which is equivalent to the Ballantrae Breccia Formation of Brookfield (1978), is at least 250 m thick and is exposed on the foreshore in the vicinity of Ballantrae Harbour. It consists of subrounded to subangular pebbles and cobbles of greywacke with a small proportion of other rock types including mudstone, chert, basalt and a variety of coarse-grained granitic and gabbroic rocks. The matrix of the breccia consists of silt and fine sand, which also occurs as rare thin discontinuous beds within the breccia.

A single lens-shaped mass of weathered dolerite about 10 m long by 1 m thick occurs within the breccia. It is concordant with the bedding in the Breccia and may be either a lava or a sill.

The Corseclays Sandstone Formation is seen in exposures on the foreshore between Bennane Lea and Balig [NX 092 838] and in the Red Burn near Corseclays Farm. It is at least 500 m thick and consists of fine-grained cross-laminated red sandstone. The grains are mainly subrounded quartz with some plagioclase. A small proportion are well rounded and probably of aeolian origin.

The Bennane Lea Breccia member is at least 10 m thick and forms the uppermost part of the Corseclays Sandstone Formation. It is exposed on the foreshore at Bennane Lea, about 3 km north of Ballantrae, and is distinguished by the appearance in the sandstone of isolated pebbles mainly of basalt with a small proportion of greywacke and quartzite. Serpentinite clasts reported by Peach and Horne (1899) during the original survey were not found during the resurvey. The pebbles range from angular to subrounded and are matrix-supported. No direct evidence for the age of the strata has been found but late Permian–early Triassic age is tentatively inferred by comparison with post-Carboniferous successions elsewhere in the south of Scotland (Smith and others, 1974; Brookfield, 1978). The dolerite lava or sill in the Park End Breccia Formation may be contemporaneous with early Permian igneous activity elsewhere in west-central Scotland.

8 Post-Ordovician igneous intrusions

Dykes belonging to several episodes of intrusion and the Ailsa Craig microgranite are present within the area of the map.

Devonian Dykes thought to be of early Devonian age have a range of orientations between north-east and west-north-west. They consist of porphyrites (plagioclase-phyric microdioritic rocks), acid porphyrites (quartz- or K-feldspar-bearing porphyrites), sparsely porphyritic felsites and, less commonly, of hornblende-lamprophyres (mainly spessartites) ranging in thickness up to 5 m. They appear to be particularly common in a zone adjacent and parallel to the Glen App Fault on its south-eastern side although this may be exposure controlled. Apart from one or two fresh examples, they are generally highly carbonated and/or sericitised, and also often carry quartz veins and cubes of pyrite.

Associated with this phase of intrusion is a variably altered hornblende-pyroxene-micadiorite mass about 0.5 km in diameter. It is very poorly exposed in the Altigabert Burn [at NX 115 776]. It is cut by a north-west-trending acid porphyrite dyke.

Permo-Carboniferous A single, north-trending, fresh alkaline lamprophyre (monchiquitic camptonite) dyke 3 m thick, crops out to the south-west of Kennedy's Pass [at NX 1464 9267]. It almost certainly belongs to the Permo-Carboniferous swarms of dykes of similar petrography that occur mainly in the west and north of Scotland, but are also known from other parts of the Southern Uplands.

Tertiary North-west-trending tholeiite and alkali olivine-dolerite (crinanite) dykes are scattered throughout the area. They are up to 2 m thick, but are usually less than 1 m thick. A few of the dykes are altered but most are fairly fresh. Vesicular margins are common. They are all probably members of the swarms associated with the Arran and Mull Tertiary igneous centres.

The island of Ailsa Craig, 15 km west of

Figure 6 Ailsa Craig, seen from the south. It consists of microgranite of Tertiary age cut by basalt dykes. (D2021)

Girvan, is an intrusion of microgranite containing the amphibole riebeckitic arfvedsonite (Howie and Walsh, 1981; Harding, 1983; Harrison and others, in press). The margins of the intrusion are not seen, but it is intruded into the Permo-Triassic strata which floor the Firth of Clyde. The microgranite is cut by numerous basic Tertiary dykes, presumably of the Arran swarm. The microgranite has been dated radiometrically at about 58.5 Ma (Macintyre, 1973). It was formerly quarried for the production of curling stone.

9 References

Bailey, E. B., and McCallien, W. J. 1952. Ballantrae igneous problems; historical review. *Trans. Edinburgh Geol. Soc.*, Vol. 15, 14–38.

Balsillie, D. 1932. The Ballantrae Igneous Complex, South Ayrshire. *Geol. Mag.*, Vol. 69, 107–131.

Balsillie, D. 1937. Further observations on the Ballantrae igneous complex, south Ayrshire. *Geol. Mag.*, Vol. 74, 20–33.

Barrett, T. J., Jenkyns, H. C., Leggett, J. K., and Robertson, A. H. F. 1982. Comment on 'Age and origin of Ballantrae ophiolite and its significance in the Caledonian orogeny and the Ordovician time scale'. *Geology*, Vol. 10, 331.

Bloxam, T. W. 1954. Rodingite from the Girvan-Ballantrae Complex, Ayrshire. *Mineral Mag.*, Vol. 30, 525–528.

Bloxam, T. W. 1955. The origin of the Girvan-Ballantrae beerbachites. *Geol. Mag.*, Vol. 92, 329–337.

Bloxam, T. W. 1968. The petrology of Byne Hill, Ayrshire. *Trans. R. Soc. Edinburgh*, Vol. 68, 105–122.

Bloxam, T. W., and Allen, J. B. 1960. Glaucophane schist, eclogite, and associated rocks from Knockormal in the Girvan-Ballantrae complex, south Ayrshire. *Trans. R. Soc. Edinburgh*, Vol. 64, 1–27.

Bluck, B. J. 1978. Geology of a continental margin: 1 The Ballantrae Complex. *In* Bowes, D. R., and Leake, B. E. (editors). *Crustal evolution in north-western Britain and adjacent regions*, 151–162. (Liverpool: Seel House.)

Bluck, B. J. 1982. Hyalotuff deltaic deposits in the Ballantrae ophiolite of SW Scotland: evidence for crustal position of the lava sequence. *Trans. R. Soc. Edinburgh, Earth Sci.*, Vol. 72, 217–228.

Bluck, B. J., Halliday, A. N., Aftalion, M., and Macintyre, R. M. 1980. Age and origin of Ballantrae ophiolite and its significance to the Caledonian orogeny and Ordovician time scale. *Geology*, Vol. 8, 492–495.

Brookfield, M. E. 1978. Revision of the stratigraphy of Permian and supposed Permian rocks of southern Scotland. *Geol. Rundschau*, Vol. 67, 110–149.

Church, W. R., and Gayer, R. A. 1973. The Ballantrae ophiolite. *Geol. Mag.*, Vol. 110, 497–592.

Cocks, L. R. M., and Toghill, P. 1973. The biostratigraphy of the Silurian rocks of the Girvan District, Scotland. *J. Geol. Soc. London*, Vol. 129, 209–243.

Hamilton, P. J., Bluck, B. J., and Halliday, A. N. 1984. Sm-Nd ages from the Ballantrae complex, SW Scotland. *Trans. R. Soc. Edinburgh: Earth Sci.*, Vol. 75, 183–187.

Harding, R. R. 1983. Zr-rich pyroxenes and glauconitic minerals in the Tertiary alkali granite of Ailsa Craig. *Scott. J. Geol.*, Vol. 19, 219–227.

Harper, D. A. T. 1984. Brachiopods from the Upper Ardmillan succession (Ordovician) of the Girvan District, Scotland. Part 1. *Monogr. Palaeontogr. Soc.*, 1–7, pls 1–11.

Harrison, R. K., Stone P., Cameron, I. B., Elliot, R. W., and Harding, R. R. In press. Geology, petrology and geochemistry of Ailsa Craig, Ayrshire. *Rep. Br. Geol. Surv.*.

Holub, F. V., Klápová, H., Bluck, B. J., and Bowes, D. R. 1984. Petrology and geochemistry of post-obduction dykes of the Ballantrae complex, SW Scotland. *Trans. R. Soc. Edinburgh: Earth Sci.*, Vol. 75, 211–223.

Howie, R. A., and Walsh, J. N. 1981. Riebeckitic arfvedsonite and aenigmatite from the Ailsa Craig microgranite. *Scott. J. Geol.*, Vol. 17, 123–128.

Ince, D. M. 1983. Shallow water facies and environments in the Ordovician of the

Girvan District, Strathclyde. PhD thesis, University of Edinburgh (unpublished).

Ingham, J. K. 1978. Geology of a continental margin: 2 Middle and late Ordovician transgression, Girvan. *In* **Bowes, D. R. and Leake, B. E.** (editors). *Crustal evolution in northwestern Britain and adjacent regions*, 163–176. (Liverpool: Seel House.)

Jelínek, E., Soucek, J., Bluck, B. J., Bowes, D. R., and Treloar, P. J. 1980. Nature and significance of beerbachites in the Ballantrae ophiolite, SW Scotland. *Trans. R. Soc. Edinburgh: Earth Sci.*, Vol. 71, 159–179.

Jones, C. M. 1977. The Ballantrae complex as compared to the ophiolites of Newfoundland. PhD thesis, University of Wales (unpublished).

Kelling, G. 1961. The stratigraphy and structure of the Ordovician rocks of the Rhinns of Galloway. *Q. J. Geol. Soc. London*, Vol. 117, 37–75.

Lapworth, C. 1882. The Girvan succession. *Q. J. Geol. Soc. London*, Vol. 38, 537–666.

Lewis, A. D. 1975. The geochemistry and geology of the Girvan-Ballantrae ophiolite and related Ordovician volcanics in the Southern Uplands of Scotland. PhD thesis, University of Wales (unpublished).

Lewis, A. D., and Bloxam, T. W. 1977. Petrotectonic environments of the Girvan-Ballantrae lavas from rare-earth element distribution. *Scott. J. Geol.*, Vol. 13, 211–222.

Lewis, A. D., and Bloxam, T. W. 1980. Basaltic macadam-breccias in the Girvan-Ballantrae Complex, Ayrshire. *Scott. J. Geol.*, Vol. 16, 181–187.

Longman, C. D., Bluck, B. J., and van Breemen, O. 1979. Ordovician conglomerates and the evolution of the Midland Valley. *Nature, London*, Vol. 280, 578–581.

Macintyre, R. M. 1973. Lower Tertiary geochronology of the north Atlantic continental margins. *In* **Pidgeon, R. T., Macintyre, R. M., Sheppard, S. M. F., and van Breemen, O.** *Geochronology and isotope geology of Scotland. Field guide and reference*, K1–K25. (European Colloquium on Geochronology III)

Peach, B. N., and Horne, J. 1899. The Silurian rocks of Britain. Vol. 1, Scotland. *Mem. Geol. Surv. UK.*

Rushton, A. W. A., and Tripp, R. P. 1979. A fossiliferous Lower Canadian (Tremadoc) boulder from the Benan Conglomerate of the Girvan district. *Scott. J. Geol.*, Vol. 15, 321–327.

Rushton, A. W. A., Stone, P., Smellie, J. L., and Tunnicliff, S. P. *In press*. An early Arenig age for the Pinbain sequence, Ballantrae Complex.

Smellie, J. L. 1984. Metamorphism of the Ballantrae Complex, south-west Scotland: A preliminary study. *Rep. Br. Geol. Surv.*, Vol. 16, No. 10, 13–17.

Smellie, J. L., and Stone, P. 1984. 'Eclogite' in the Ballantrae Complex: a garnet clinopyroxenite segregation in mantle harzburgite? *Scott. J. Geol.*, Vol. 20, 315–327.

Smith, D. B., Brunstrom, R. G. W., Manning, P. I., Simpson, S., and Shotton, F. W. 1974. A correlation of Permian rocks in the British Isles. *Spec. Rep. Geol. Soc. London*, No. 5, 1–45.

Spray, J. G., and Williams, G. D. 1980. The sub-ophiolite metamorphic rocks of the Ballantrae igneous complex, SW Scotland. *J. Geol. Soc.*, Vol. 137, 359–368.

Stone, P. 1984. Constraints on genetic models for the Ballantrae complex, SW Scotland. *Trans. R. Soc. Edinburgh: Earth Sci.*, Vol. 75, 189–191.

Stone, P., Gunn, A. G., Coats, J. S., and Carruthers, R. M. *In press*. In **Jones, M. V.** (editor). *Mineral exploration in the Ordovician Ballantrae Complex, SW Scotland.* (London: Institution of Mining and Metallurgy.)

Stone, P., Lambert, J. L. M., Carruthers, R. M., and Smellie, J. L. 1984. Concealed ultramafic bodies in the Ballantrae Complex, south-west Scotland: borehole results. *Rep. Br. Geol. Surv.*, Vol. 16, No. 1, 41–45.

Stone, P., and Rushton, A. W. A. 1983. Graptolite faunas from the Ballantrae ophiolite complex and their structural implications. *Scott. J. Geol.*, Vol. 19, 297–310.

Stone, P., and Smellie, J. L. 1986. Ballantrae. Description of the solid geology of parts of 1:25 000 sheets NK 08, 09, 18 and 19. Classical areas of British Geology. British Geological Survey. (London: HMSO.)

Stone, P., and Strachan, I. 1981. A fossiliferous borehole section within the Ballantrae ophiolite *Nature, London*, Vol. 293, 455–456.

Thirwall, M. F., and Bluck, B. J. 1984. Sr-Nd isotope and geochemical evidence that the Ballantrae 'ophiolite', SW Scotland, is polygenetic. *In* **Gass, I. G., Lippard, S. J., and Shelton, A. W.** (editors). Ophiolites and oceanic lithosphere, 215–230. *Spec. Rep. Geol. Soc. London*, No. 13.

Treloar, P. J., Bluck, B. J., Bowes, D. R., and Dudek, A. 1980. Hornblende-garnet metapyroxenite beneath serpentinite in the Ballantrae complex of SW Scotland and its bearing on the depth provenance of obducted oceanic lithophere. *Trans. R. Soc. Edinburgh: Earth Sci.*, Vol. 71, 201–212.

Tripp, R. P. 1962. Trilobites from the *confinis* Flags (Ordovician) of the Girvan District, Ayrshire. *Trans. R. Soc. Edinburgh*, Vol. 65, 1–40.

Tripp, R. P. 1976. Trilobites from the basal *superstes* Mudstones (Ordovician) at Aldons Quarry, near Girvan, Ayrshire. *Trans. R. Soc. Edinburgh*, Vol. 69, 369–423.

Tripp, R. P. 1979. Trilobites from the Ordovician Auchensoul and Stinchar Limestones of the Girvan District, Strathclyde. *Palaeontology*, Vol. 22, 339–361.

Tripp, R. P. 1980. Trilobites from the Ordovician Balclatchie and lower Ardwell groups of the Girvan district, Scotland. *Trans. R. Soc. Edinburgh, Earth Sci.*, Vol. 71, 123–145.

Tripp, R. P., Williams, A., and Paul, C. R. C. 1981. On an exposure of the Ordovician *superstes* Mudstones at Colmonell, Girvan District, Strathclyde. *Scott. J. Geol.*, Vol. 17, 21–25.

Tyrrell, G. W. 1909. A new occurrence of picrite in the Ballantrae district and its associated rocks. *Trans. Geol. Soc. Glasgow*, Vol. 13, 283–290.

Walton, E. K. 1957. Two Ordovician conglomerates in South Ayrshire. *Trans. Geol. Soc. Glasgow*, Vol. 22, 133–156.

Walton, E. K. 1961. Some aspects of the succession and structure in the lower Palaeozoic rocks of the Southern Uplands of Scotland. *Geol. Rundsch.*, Vol. 50, 63–77.

Wilkinson, J. M., and Cann, J. R. 1974. Trace elements and tectonic relationships of basaltic rocks in the Ballantrae igneous complex, Ayrshire. *Geol. Mag.*, Vol. 111, 35–41.

Williams, A. 1959. A structural history of the Girvan district, SW Ayrshire. *Trans. R. Soc. Edinburgh*, Vol. 63, 629–667.

Williams, A. 1962. The Barr and Lower Ardmillan Series (Caradoc) of the Girvan district, south-west Ayrshire, with descriptions of the brachiopoda. *Mem. Geol. Soc. London*, No. 3, 1–267.

Williams, G. D., and Spray, J. G. 1979. Non-cylindrical, flexural slip folding in the Ardwell Flags: a statistical approach. *Tectonophysics*, Vol. 58, 269–277.

10 Glossary

Accretionary prism A complex of steeply inclined strata formed by the sequential stacking, by underthrusting, of successively younger packets of stratified rocks against a continental margin adjacent to a subduction zone.
Acid Relating to igneous rocks containing more than 63 per cent of silica.
Acid porphyrite An intrusive igneous rock with phenocrysts of acid plagioclase in a groundmass of alkali and sodic feldspar, sparse mafic minerals and quartz; usually very altered.
Allochthonous Formed elsewhere and transported to its present site.
Amphibolite A rock consisting mainly of amphibole and plagioclase with little or no quartz.
Amygdaloidal Descriptive of a lava containing gas cavities or vesicles which are filled with secondary minerals.
Autoclastic Having a broken or brecciated condition which formed in place.
Basic Relating to igneous rocks containing less than 52 per cent of silica.
Beerbachite A thermally metamorphosed rock with a granoblastic texture and of doleritic composition.
Brachiopods Marine animals that were abundant and diverse in the Palaeozoic, characterised by two dissimilar but symmetrical shells.
Breccia A coarse-grained clastic rock composed of angular rock fragments set in a finer-grained matrix.
Clinopyroxenite An ultramafic plutonic rock composed essentially of clinopyroxene.
Cornstone Calcareous concretions forming either a zone of nodules or a continuous layer.
Cumulate An igneous rock formed by the accumulation of crystals concentrated from a magma by gravity.
Desilication The removal of silica from a rock magma by the breakdown of silicates.
Diorite A coarse-grained intermediate igneous rock.
Dolerite A medium-grained basic igneous rock.
Dunite An ultramafic rock composed essentially of olivine.
Dynamothermal metamorphism Metamorphism produced by the combined effects of confining pressure, shear stress and heat.
Eclogite A plutonic rock composed essentially of garnet and the green pyroxene omphacite.
Felsite A pale, commonly pink, fine-grained compact rock, composed of alkali feldspar, quartz and sparse iron ore.
Flaser A streaky, patchy structure developed in a granular igneous rock as a result of dynamometamorphism.
Gabbro A coarse-grained basic igneous rock.
Granoblastic A texture in metamorphic rock in which recrystallisation formed essentially equidimensional polygonal crystals.
Graptolites An extinct group of colonial marine animals that lived in small 'chitinous' tubes and floated or swam in the open sea. They are particularly useful for assessing the ages of the sedimentary rocks in which they occur.
Greywacke A poorly sorted sandstone with angular to subangular mineral and rock fragments set in a clayey matrix.
Harzburgite An ultramafic rock composed mainly of olivine and orthopyroxene.
Hornfels A hard tough fine-grained rock produced by thermal metamorphism.
Hyaloclastite A deposit composed of comminuted basaltic glass formed by the fragmentation of glassy lava or by the violent eruption of basalt under water.
Intra-plate Within the margins of a lithospheric plate.
Island arc An arcuate chain of islands associated, on its convex side, with a deep ocean trench which marks the site of a subduction zone.

Lamprophyre A dark coloured porphyritic dyke rock rich in phenocrysts of biotite or other mafic minerals.
Mélange A body of rock characterised by the inclusion of fragments and blocks of all sizes, both exotic and native, in a sheared matrix.
Metamorphism A change in the character of a rock in response to an alteration in physical or chemical conditions.
Metasomatism A change in the composition of a rock by the removal or introduction of chemical constituents.
Microgranite A medium-grained acid igneous rock similar in composition to granite.
Oligomict Consisting of fragments of only one rock-type.
Olistostrome A deposit consisting of a chaotic unbedded mass of heterogeneous sediments and/or rock-types that accumulated by submarine sliding or slumping.
Ophiolite An association of basic and ultrabasic rocks such as spilite, gabbro and peridotite and which commonly has suffered serpentinisation.
Pegmatitic Having a very coarse-grained texture.
Phenocrysts Relatively large crystals in the fine-grained groundmass of an igneous rock.
Pillow lava Lava which has been extruded under water and has the form of a pile of pillow-like masses.
Polymict Consisting of fragments of many different rock-types.
Porphyrite A compact fine-grained intrusive igneous rock, with phenocrysts of intermediate plagioclase (usually albitised), hornblende, pyroxene or biotite in a matrix of these minerals and some iron ore, usually highly altered.
Porphyritic Descriptive of igneous rocks containing relatively large crystals (phenocrysts) set in a finer-grained matrix.
Porphyroclast Relict crystals or fragments of crystals in a metamorphic rock which are set in a finer-grained crushed or recrystallised matrix.

Pyroclastic Consisting of fragmental volcanic material thrown out of a volcano.
Pyroxenite An ultrabasic rock consisting essentially of pyroxene.
Plutonic Descriptive of an igneous rock which crystallised at depth.
Rodingite A gabbroic rock which has suffered calcium metasomatism and now consists of calcium-rich secondary minerals, such as grossular and prehnite.
Schist A metamorphic rock with a distinct parallel arrangement of the constituent minerals.
Serpentinite An altered ultrabasic rock consisting of the mineral serpentine, which is an alteration product of olivine and pyroxene.
Spilite An altered basalt in which the feldspar has been albitised and the dark (mafic) minerals altered to low temperature hydrous minerals.
Spreading ridge A mid-oceanic ridge along which two lithospheric plates are moving apart, and which is the locus of the formation of oceanic crust.
Subduction The process by which one lithospheric plate slips down below another at a converging plate boundary.
Tholeiite A type of basalt consisting of plagioclase and pigeonite together with interstitial glass or quartz-alkali feldspar intergrowths.
Trilobites An extinct group of marine arthropods with a segmented body. Some swam in the sea but most lived on the sea floor.
Trondhjemite A light-coloured plutonic rock consisting mainly of sodic plagioclase and quartz, and practically devoid of dark (mafic) minerals.
Tuffaceous Descriptive of a rock containing a proportion of volcanic ash.
Turbidite The lithified sediment deposited from a turbidity current, commonly as an argillaceous sandstone, but varying in grade from mudstone to conglomerate.
Ultramafic Applied to igneous rocks consisting mainly of ferromagnesian minerals (olivine, pyroxene, etc) and almost devoid of quartz and feldspar.

Vesicular Containing small cavities formed by gas bubbles in volcanic rocks.
Wehrlite An ultrabasic rock consisting of olivine with lesser amounts of augite.
Xenolith A fragment of a pre-existing rock within an igneous rock.
Zone A thickness of strata characterised by a particular and unique fossil assemblage. Zones are used to assess the relative ages of rock-sequences.

11 BGS publications relevant to the area

Maps at 1:50 000 scale
Girvan (Sheet 7) Solid edition. *In press.*
Girvan (Sheet 7) Drift edition. 1980.

Maps at 1:25 000 scale
Ballantrae (parts of sheets NX 08, 18 and 19). Solid editions. *In press.*
Ailsa Craig (parts of sheets NS 00 and NX 09). Solid edition. *In press.*

Maps at 1:10 000 scale Solid editions
NX 07 NE (1984)
NX 07 SW (in part) (1984)
NX 08 NE (1984)
NX 08 SE (1984)
NX 17 NW (1984)
NX 18 NE (1984)
NX 18 NW (1984)
NX 18 SE (1984)
NX 18 SW (1984)
NX 19 NE (1984)
NX 19 SE (1984)
NX 19 SW (1984)

Reports of BGS and IGS
Vol. 16, No. 1 Accretionary lapilli and highly vesiculated pumice in the Ballantrae ophiolite complex: ash-fall products of subaerial eruptions. J. L. Smellie, 1984.

Vol. 16, No. 10 Metamorphism of the Ballantrae Complex, SW Scotland: a preliminary review. J. L. Smellie, 1984.

No. 82/1 Clastic rocks within the Ballantrae Complex: borehole evidence. P. Stone, 1982.

Vol. 16, No. 1 Concealed ultramafic bodies in the Ballantrae Complex, SW Scotland: borehole results. P. Stone, J. L. M. Lambert, R. M. Carruthers and J. L. Smellie, 1984

Ballantrae. Description of the solid geology of parts of 1:25 000 sheets NX 08. 09 18 and 19. Classical area of British Geology. P. Stone and J. L. Smellie, 1986

In press. Geology and geochemistry of Ailsa Craig, Ayrshire. R. K. Harrison, P. Stone, I. B. Cameron, R. W. Elliot, and R. R. Harding.